PAT: Portable Appliance Testing

To Carolyn

PAT: Portable Appliance Testing

In-service inspection and testing of electrical equipment

Brian Scaddan IEng, MIIE(elec)

Newnes

OXFORD AUCKLAND BOSTON JOHANNESBURG MELBOURNE NEW DELHI

Newnes
An imprint of Butterworth-Heinemann
Linacre House, Jordan Hill, Oxford OX2 8DP
225 Wildwood Avenue, Woburn, MA 01801-2041
A division of Reed Educational and Professional Publishing Ltd

 A member of the Reed Elsevier plc group

First published 2000

British Library Cataloguing in Publication Data
A catalogue record for this book is available from the British Library

ISBN 0 7506 4721 3

Typeset by David Gregson Associates, Beccles, Suffolk
Printed and bound in Great Britain by Biddles Ltd
www.biddles.co.uk

PLANT A TREE

BTCV
*British Trust for
Conservation Volunteers*

FOR EVERY TITLE THAT WE PUBLISH, BUTTERWORTH-HEINEMANN
WILL PAY FOR BTCV TO PLANT AND CARE FOR A TREE.

Contents

Preface

The introduction of The Electricity at Work Regulations (EAWR)1989, prompted, among many other things, a rush to inspect and test *portable appliances*. The Regulations do not require such inspecting and testing, nor do they specifically mention portable appliances. They do, however, require any electrical system to be constructed, maintained and used in such a manner as to prevent danger, and in consequence inspection and testing of systems (portable appliances are systems) is needed in order to determine if maintenance is required.

All electrical equipment connected to the fixed wiring of an installation will need attention, not just portable appliances. I have however left the title of this book as 'Portable Appliance Testing' as such words are now indelibly imprinted on our minds, even though it should read 'Inspection and Testing of In-service Electrical Equipment'.

The book is intended for those who need be involved in this inspection and testing process, either as a business venture or as an 'in-house' procedure to conform with the EAWR. It is also a useful reference document for anyone embarking on a City & Guilds 2377 course.

Brian Scaddan

1
Legislation

There are four main sets of legislation that are applicable to the inspection and testing of in-service electrical equipment:

- The Health & Safety at Work Act 1974 (H&SWA)
- The Management of the H&SWA Regulations 1992
- The Provision & Use of Work Equipment Regulations 1992
- The Electricity at Work Regulations 1989 (EAWR)

The Health & Safety at Work Act 1974

This applies to all persons – employers and employees – at work, and places a duty of care on all to ensure the safety of themselves and others.

The Management of the Health & Safety at Work Act Regulations 1992

In order that the H&SWA can be effectively implemented in the workplace, every employer has to carry out a risk assessment to ensure that employees and those not in his/her employ, are not subjected to danger.

The Provision & Use of Work Equipment Regulations 1992

Work equipment must be constructed in such a way that it is suitable for the purpose for which it is to be used. Once again, the employer is responsible for these arrangements.

The Electricity at Work Regulations 1989

These regulations, in particular, are very relevant to the inspection and testing of in-service electrical equipment.

There are two important definitions in the EAWR: the electrical system and the duty holder.

Electrical system

This is anything that generates, stores, transmits or uses electrical energy, from a power station to a wrist-watch battery. The latter would not give a person an electric shock, but could explode if heated, giving rise to possible injury from burns.

Duty holder

This is anyone (employer, employee, self-employed person etc.) who has 'control' of an electrical system. Control in this sense means designing, installing, working with or maintaining such systems. Duty holders have a legal responsibility to ensure their own safety and the safety of others whilst in control of an electrical system.

The EAWR does not specifically mention inspection and testing; it simply requires electrical systems to be 'maintained' in a condition so as not to cause danger. However, we only know if a system needs to be maintained if it is inspected and tested, and thus the need for such inspection and testing of a system is implicit in the requirement for it to be maintained.

Anyone who inspects and tests an electrical system is, in law, a duty holder and must be competent to undertake such work.

Prosecutions

Here are just a few examples of the many prosecutions under the Electricity at Work Regulations 1989 that take place every year.

Case 1

A greengrocer was visited, probably for the second time, by the Health and Safety Executive inspectors, who found eleven faults with the electrical installation. They were:

1 a broken fuse to a fused connection unit;
2 a broken three-way lighting switch;
3 a broken double socket outlet;
4 a broken bayonet light fitting;
5 a missing ceiling rose cover;
6 the flexible cord feeding the beetroot boiler went under the casing and not through the proper hole in the side;
7 there was no earthing to a fluorescent fitting;
8 there was no earthing to a metal spotlight;
9 block connectors were used to connect some bulkhead lights;

10   block connectors were used to connect the fluorescent lights;

11   block connectors were used to connect a spotlight.

He was subsequently fined £4,950, and although he was 'only a greengrocer', he was also a duty holder, and as such had a responsibility for the safety of the staff working in the shop.

Case 2

An electrician received serious burns to his face, arms and legs after he was engulfed in a ball of flames whilst testing an old motor control switch-board. He was reaching into the board to test contacts located only a few inches away from exposed, live, 400 V terminals when the accident happened. He was apparently using inappropriate test leads that were un-fused and had too much exposed metal on the tips. He was also working near live terminals because no arrangement had been made for the board to be made dead.

His company was fined a total of £1,933 because they did not prevent work on or near live equipment. They were duty holders. The electrician, however, also a duty holder, carried the main responsibility for the accident, but would not have been prosecuted, as he was the only one to be injured.

Case 3

A young foreman on a large construction site was electro-cuted when he touched the metal handle of a site hut which had become live. An employee of the company carrying out the electrical contracting work on the site had laid inadequate wiring in the hut which had later been crushed by its weight, causing a fault. Consequently the residual current device (r.c.d.) protecting the hut kept tripping out, as it should have. However, another of the electrical contractor's employees by-passed the r.c.d. so that it would not trip. This caused the site hut to become live.

The construction company was fined £97,000 for failing to monitor site safety, the electrical contractors were fined £30,000, and the contractor's managing director was fined £5,000 and disqualified from being a company director for three years.

2
Setting up

There are two ways for an organisation to ensure that in-service electrical equipment is regularly maintained:

- employ a specialist company to provide the inspection and testing service, or
- arrange for trained and competent 'in-house' staff to carry out the work.

In either case, the first step is for the organisation to appoint a 'responsible person' – who will, therefore, be a duty holder – to whom staff and/or outside contractors should report the results of any inspection and test, including defects etc. Such a person could be the manager of the premises or a member of staff: they will need to be trained and competent, both in the management of the appliance testing process and in the knowledge of relevant legislation as discussed in Chapter 1.

The second step is for the 'responsible person' to carry out an inventory of all equipment that may need testing and/or inspecting, and make decisions as to the frequency of such work. Some advice may be needed here from an experienced contractor in order to achieve the most effective time schedule and to make decisions on which equipment should be involved.

Table 2.1 gives some examples of recommended periods between each inspection and test.

The 'responsible person' should have in place a procedure for users of electrical equipment to report and log any defects found.

Whether the inspection and test is to be carried out by competent staff or by outside contractors, it is advisable that various forms be produced to record:

- details of equipment that may need to be inspected and tested (Form 1)
- the results of formal visual inspection or combined inspection and testing (Form 2)
- details of faulty equipment taken out of service and sent for repair (Form 3)

Previous records must be kept and made available to any person conducting routine inspection and testing of in-service electrical equipment.

Form 1

EQUIPMENT REGISTER						
COMPANY: *Jones Footware Ltd., Blacktown.*						
					Frequency of Insp. & Test	
Register No.	Equipment	Equip. No. *	Class I, II or III	Normal Location	Formal visual Insp.	Combined Insp.& Test
1	*Kettle*	*12*	*I*	*Kitchen*	*6 mths.*	*12 mths.*
2						
3						
4						
5						
6						
7						

* This could be the serial No. or a number allocated by the company or the contractor and durably marked on the equipment

Table 2.1. Sample of suggested frequencies of inspection and testing

Equipment	Class	Inspection and tests	Offices and shops	Hotels	Schools
Handheld	Class I and II	User checks	Before use	Before use	Before use
	Class I	Formal visual insp.	Every 6 months	Every 6 months	Every 4 months
		Combined insp. and test	Every year	Every year	Every year
	Class II	Formal visual insp.	Every 6 months	Every 6 months	Every 4 months
		Combined insp. and test	None	None	Every 4 years
Portable	Class I and II	User checks	Weekly	Weekly	Weekly
	Class I	Formal visual insp.	Every year	Every year	Every 4 months
		Combined insp. and test	Every 2 years	Every 2 years	Every year
	Class II	Formal visual insp.	Every 2 years	Every 2 years	Every 4 months
		Combined insp. and test	None	None	Every 4 years
Moveable	Class I and II	User checks	Weekly	Weekly	Weekly
	Class I	Formal visual insp.	Every year	Every year	Every 4 months
		Combined insp. and test	Every 2 years	Every 2 years	Every year
	Class II	Formal visual insp.	Every 2 years	Every 2 years	Every 4 months
		Combined insp. and test	Every 2 years	Every 2 years	Every 4 years
Stationary	Class I and II	User checks	None	None	Weekly
	Class I	Formal visual insp.	None	None	None
		Combined insp. and test	Every 4 years	Every 4 years	Every year
	Class II	Formal visual insp.	Every 2 years	Every 2 years	Every year
		Combined insp. and test	None	None	Every 4 years
IT	Class I and II	User checks	None	None	Weekly
	Class I	Formal visual insp.	Every 2 years	Every 2 years	None
		Combined insp. and test	Every 4 years	Every 4 years	Every year
	Class II	Formal visual insp.	Every 2 years	Every 2 years	Every year
		Combined insp. and test	None	None	Every 4 years

Form 2

INSPECTION and TESTING RECORD

COMPANY: _R.F Bloggins & Son Ltd., Whiteford_

Equipment	Equip. No.	Class I, II or III	Normal Location
Floor polisher	8	1	Store room

Make :	Lynatron	Voltage :	230	V	Purchase date :	1.2.99
Model :	KPX 2	Power :	700	W		
Serial No. :	13579	Current :	N/A	A		
		Fuse :	5	A		

Frequency of inspection and testing		
Formal visual	Combined insp. & test	
Weekly	12 mths	

	Inspection							Testing						
								Earth continuity		Insulation resistance				
Date	Correct environment for use	Permission to disconnect *	Socket	Plug	Flex	Body		Ohms	OK	App. Voltage M-ohms	E. Leakage mA	Functional	OK to use	Signature
1.2.2000	Yes	N/A	OK	OK	OK	OK		0.07	Yes	200+	N/A	OK	YES	A. Mann
8.2.2000	Yes	N/A	OK	OK	OK	OK							YES	A. Mann
15.2.2000	Yes	N/A	OK	OK	OK	OK							YES	A. Mann
22.2.2000	Yes	N/A	OK	OK	OK	OK							YES	A. Mann

* Applies to business and IT equipment which may need downloading first

10

Form 3

FAULTY EQUIPMENT & REPAIR REGISTER

COMPANY: _Mr. Baldys Hairdressing Emporium , Thintown_

Date removed from service	Equipment	Equip. No.	Equipment register No.	Normal location	Fault	Date sent for repair	Repairer	Date returned	Suitable for use OK	Signature	Comments
13.3.2000	Hair dryer	9	4	Main salon	Frayed flex	20.3.2000	N.O.Good	28.3.2000	Yes	N.O. Good	
15.3.2000	Curling tongs	11	18	Room 2	Cracked handle	20.3.2000	T.O. Bad	1.4.2000	No	T.O.Bad	Not repairable

3
Equipment to be inspected and tested

As mentioned in the introduction to this book, it is not just portable appliances that have to be inspected and tested, but all in-service electrical equipment. This includes items connected to the supply by 13 A BS1363 plugs, BS EN 60309-2 industrial plugs or hard wired to the fixed installation via fused connection units or single or three-phase isolators.

It is perhaps wise at this stage to comment on the two methods of receiving an electric shock – direct and indirect contact – and the different classes of equipment (Class 0, Class 01, Class I, Class II and Class III).

Direct contact

This is touching intentionally live parts. Protection is generally achieved by applying basic insulation to such parts and/or enclosing them to prevent contact.

Indirect contact

This is contact with exposed metalwork of electrical equipment that has become live due to a fault, e.g. breakdown of basic insulation. Protection is generally by adequate earthing or the use of double or reinforced insulation (Class II).

Class 0 equipment or appliances

Almost everyone can remember those old-fashioned, ornate brass table lamps, wired with either flat PVC-insulated twin flex or twisted cotton-covered rubber-insulated twin flex. In other words, equipment with a non-earthed metal case, the protection against electric shock being provided by insulating live parts with basic insulation only. Breakdown of this insulation could result in the metal enclosure becoming live and with no means of disconnecting the fault.

Class 01 equipment or appliances

This is the same as Class 0. However, the metal casing has an earthing terminal but the supply cable is twin and the plug has no earth pin.

Class 0 and 01 equipment may be used but only in special circumstances and in a strictly controlled environment. Generally these classes should not be used unless connections to earth are provided on the item and an earth return path via a supply cable that has a circuit protective conductor (c.p.c.) incorporated: this would convert the equipment to Class I.

Class I equipment or appliances

These items have live parts protected by basic insulation and a metal enclosure or accessible metal parts that could become live in the event of failure of the basic insulation (indirect contact). Protection against shock is by basic

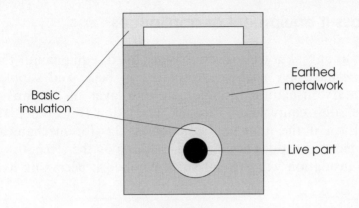

Earthed
metalwork

Basic
insulation

Live part

Figure 3.1

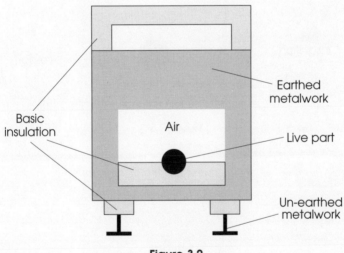

Earthed
metalwork

Basic
insulation

Air

Live part

Un-earthed
metalwork

Figure 3.2

insulation and earthing via casing the c.p.c. in the supply
cable and the fixed wiring.

Typical Class I items include toasters, kettles, washing
machines, lathes and pillar drills (see Figs 3.1–3.2).

Class II equipment or appliances

Commonly known as double-insulated equipment, the items have live parts encapsulated in basic and supplementary insulation (double), or one layer of reinforced insulation equivalent to double insulation (Figs 3.3–3.4).

Even if the item has a metal casing (for mechanical protection) it does not require earthing as the strength of the insulation will prevent such metalwork becoming live

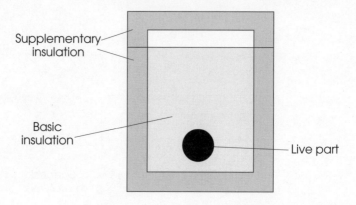

Supplementary insulation

Basic insulation

Live part

Figure 3.3

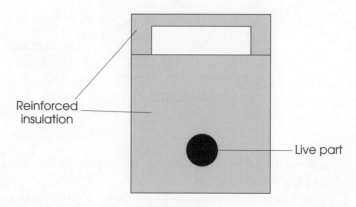

Reinforced insulation

Live part

Figure 3.4

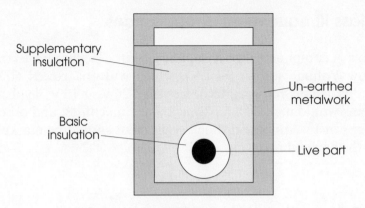

Figure 3.5

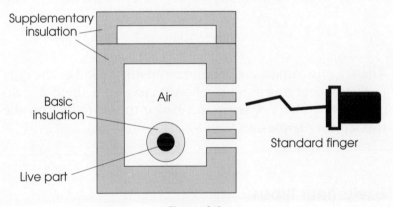

Figure 3.6

under fault conditions. The cable supplying such equipment will normally be two core with no c.p.c. (Fig. 3.5).

Examples of Class II equipment would include most modern garden tools such as hedge trimmers and lawn mowers and also food mixers, drills table lamps etc. All such items should display the Class II equipment symbol:

Equipment with grlls or openings, e.g. hair dryers, need to pass the standard finger entry test (Fig. 3.6).

Class III equipment or appliances

This is equipment that is supplied from a Separated Extra Low Voltage source (SELV), which will not exceed 50 V and is usually required to be less than 24 V or 12 V. Typical items would include telephone answer machines, and other items of IT equipment. Such equipment should be marked with the symbol:

III

and be supplied from a safety isolating transformer to BS3535 which in itself should be marked with the symbol:

These transformers are common and are typical of the type used for charging mobile phones etc. Note there are no earths in a SELV system and hence the earth pin on the transformer is plastic.

Equipment types

The Code of Practice defines various type of equipment/accessory that needs to be inspected and tested and that are generally in normal use. Advice from the manufacturer should be sought before testing specialist equipment. The equipment types are as follows:

- portable equipment/appliances
- hand held equipment/appliances
- moveable equipment/appliances
- stationary equipment/appliances
- fixed equipment/appliances
- built-in equipment/appliances

- information technology equipment
- extension leads.

Portable equipment/appliances

These are items which are capable of easy movement whilst energised and/or in operation. Examples of such appliances are:

- chip fryers
- toasters
- coffee percolators
- tin openers.

Hand held equipment/appliances

These items are of a portable nature that require control/ use by direct hand contact. Examples include:

- drills
- hair dryers
- hedge trimmers
- soldering irons.

Moveable equipment/appliances

There is a thin dividing line between this and the previous two types, but in any case still needs inspecting and testing. Generally such items are 18 kg or less and have wheels or are easily moved. Examples would include:

- Tumble dryers
- the old fashioned twin-tub washing machine
- industrial/commercial kitchen equipment.

Stationary equipment/appliances

This is equipment in excess of 18 kg and which is not intended to be moved about, such as:

- ordinary cookers
- dishwashers
- washing machines.

Fixed equipment/appliances

These items are fixed or secured in place, typically:

- tubular heaters
- lathes and other industrial equipment
- towel rails.

Built-in equipment

This is equipment that is 'built-in' to a unit or recess such as:

- an oven
- an inset electric fire.

Information technology (IT) equipment

In general terms, this is business equipment such as:

- PCs
- printers
- typewriters
- scanners.

Extension leads

These include the multi-way sockets so very often used where IT equipment is present, as there is seldom enough fixed socket outlets to supply all the various units. These leads should always be wired with 3 core (phase neutral and earth) cable, and should not exceed:

- 12 m in length for a 1.25 mm^2 core size
- 15 m in length for a 1.5 mm^2 core size
- 25 m in length for a 2.5 mm^2 core size.

The latter should be supplied via a **BS EN** 60309-2 plug, and if any of the lengths are exceeded, the leads should be protected by a **BS**7071 30 mA r.c.d.

4
Inspection

Inspection is vital, and must precede testing. It may reveal serious defects which may not be detected by testing only.

Two types of inspection are required: user checks and formal visual inspection.

User checks

All employees are required by the Electricity at Work Regulations to work safely with electrical appliances/ equipment and hence all should receive some basic training/instruction in the checking of equipment before use. (This training need only be of a short duration.) Generally, this is all common sense: nevertheless, a set routine or pre-use checks should be established. Such a routine could be as follows:

- Check the condition of the appliance/equipment (look for cracks or damage).
- Examine the cable supplying the item, looking for cuts, abrasions cracks etc.
- Check the cable sheath is secure in the plug and the appliance.
- Look for signs of overheating.

- Check that it has a valid label indicating that it has been formally inspected and tested.
- Decide if the item is suitable for the environment in which it is to be used, e.g. 240 V appliances should not be used on a construction site.
- If all these checks prove satisfactory, check that the appliance is working correctly.

If the user feels that the equipment is not satisfactory, it must be switched off, removed from the supply, labelled 'Not to be used' or words to that effect, and reported to a responsible person. That person will then take the necessary action to record the faulty item and arrange remedial work or have it disposed of.

No record of user checks is required if the equipment is considered satisfactory.

Formal visual inspection

This must be carried out by a person competent to do so, and recorded on an appropriate form. This inspection is similar to, but more detailed than, user checks and must be conducted with the accessory/equipment disconnected from the supply.

General

- Check cable runs to ensure that cables will not be damaged by staff or heavy equipment.
- Make sure that plugs, sockets, flex outlets, isolators etc. are always accessible to enable disconnection/isolation of the supply, either for functional, maintenance or emergency purposes. For example, in many office

environments, socket outlets are very often obscured by filing cabinets etc.

- Check that items that require clear ventilation such as convector heaters, VDU's etc. are not covered in paper, files etc. and that foreign bodies or moisture cannot accidentally enter such equipment.
- Ensure that cables exiting from plugs or equipment are not tightly bent.
- Check that multiway adaptors/extension leads are not excessively used.
- Check that equipment is suitable for both the purpose to which it is being put and the environment in which it is being used.
- Ensure that accessories/equipment are disconnected from the supply during the inspection process, either by removing the plug or by switching off at a connection unit or isolator.
- Take great care before isolating or switching off business equipment. Ensure that a responsible person agrees that this may be done, otherwise this may result in a serious loss of information, working processes etc.

The accessories/equipment

- Check the cable for damage. Is it too long or too short?
- Is the supply cable/cord to the appliance the right size?
- Is the plug damaged? Look for signs of overheating etc.
- Is the fuse in a BS1363 13 plug the correct size? Are the contacts for the fuse secure? (This requires dismantling of the plug. The fuse should be approved, and ideally have an ASTA mark on it. Some fuses made in China and marked PMS are dangerous and should be replaced. Fuse and cord sizes (in accordance with BS1363) in relation to appliance rating are generally:

Appliance rating	Fuse size	Cord size
Up to 700 W	3 A	$0.5\,\text{mm}^2$
700–3000 W	13 A	$0.75–1.55\,\text{mm}^2$

- If a plug is damaged and is to be replaced, ensure that the replacement has sleeved live pins. The Plugs and Sockets etc. Regulations 1994 makes it illegal to sell plugs without such sleeved pins. However, this requirement is not retrospective, in that it does apply to plugs with un-sleeved pins already in use.

5
Combined inspection and testing

Combined inspection and testing comprises preliminary inspection as per Chapter 4 together with instrument tests to verify earth continuity, insulation resistance, functional checks and, in the case of cord sets and extension leads, polarity as well. In some low-risk areas such as offices, shops, hotels etc., Class II equipment does not require the routine instrument tests.

Testing

This has to be carried out with the appliance/equipment isolated from the supply. Such isolation is, of course, easy when the item is supplied via a plug and socket, but presents some difficulties if it is permanently wired to, say, a flex outlet, a connection unit, or an isolator etc. In these cases the tester must be competent to undertake a disconnection of the appliance, if not, then a qualified/competent electrical operative should carry out the work.

Additionally, the permission of a responsible person may be needed before isolating/disconnecting business equipment.

Preliminary inspection

This must always be done before testing as it could reveal faults that testing may not show, such as unsecured cables in appliance housings, damaged cable sheathing etc. The inspection procedure is as detailed in Chapter 4.

Testing

This may be carried out using a portable appliance tester, of which there are many varieties, or separate instruments capable of measuring continuity and insulation resistance.

Portable appliance testers

These instruments allow appliances, fitted with a plug, to be easily tested. Some testers have the facility for testing appliances of various voltage ranges, single and three phase, although the majority only accept single phase 240 V or 110 V plugs (BS1363 and BS EN 60309-2)

Generally, portable appliance testers are designed to allow operatives to 'plug in' an item of equipment, push a test button, view results and note a 'pass' or 'fail' indication. The operative can then interpret these results and, where possible, make adjustments which may enable a 'fail' indication to be changed to a 'pass' status.

Some portable appliance testers are of the GO, NO-GO type where the indication is either a red (fail) or green (pass) light. As there are no test figures associated with this type of tester, no adjustment can be made. This could result in appliances being rejected when no fault is present. This situation will be dealt with a little later.

Continuity/insulation resistance testers

These are usually dual instrument testers, although separate instruments are in use. Multimeters are rarely suitable for these tests.

For earth continuity, the instrument test current (a.c. or d.c.) should be between 20 and 200 mA with the source having an open circuit voltage of between 100 mV and 24 V. For insulation resistance the instrument should deliver a maintainable test voltage of 500 V d.c. across the load. **Note: All test leads should conform to the recommendations of the HSE Guidance Note GS 38.**

So, what are the details of the tests required?

Earth continuity

This test can only be applied to Class I equipment, and the purpose of the test is to ensure that the earth terminal of the item is connected to the casing effectively enough to result in the test between this terminal and the casing giving a value of not more than 0.1 Ω.

Clearly, it is not very practicable to have to access terminals inside an enclosure and hence, it is reasonable to measure the earth continuity from outside, via the plug and supply lead. This also checks the integrity of the lead earth conductor, or c.p.c.

Testing in this way will, of course, add the resistance of the lead to the appliance earth resistance, which could result in an overall value in excess of the 0.1 Ω limit, and the tester may indicate a 'fail' status. This is where the interpretation of results is so important in that, provided the final value having subtracted the lead resistance from the instrument reading, is no more than 0.1 Ω, the appliance can be passed as satisfactory.

The use of a GO, NO-GO instrument prohibits such an adjustment as there are no test values available.

The following table gives the resistance in ohms per metre of copper conductors, at 20°C for flexible cords from $0.5\,\text{mm}^2$ to $4.0\,\text{mm}^2$.

Conductor size (mm^2)	Resistance (Ω/m)
0.5	0.039
0.75	0.026
1.0	0.0195
1.25	0.0156
1.5	0.013
2.5	0.008
4.0	0.005

Hence the c.p.c. of 5 m of $1.0\,\text{mm}^2$ flexible cord would have a resistance of:

$$5 \times 0.0195 = 0.0975\ \Omega$$

It is unlikely that appliances in general use will have supply cords in excess of $1.25\,\text{mm}^2$ as the current rating for such a cord is 13 A, which is the maximum for a BS1363 plug.

Example

The measured value of earth continuity for an industrial floor polisher, using a portable appliance tester, is $0.34\ \Omega$. The supply cord is 10 m long and has a conductor size of $0.75\,\text{mm}^2$. The test instrument also indicates a 'fail' condition. Can the result be overruled?

Resistance of c.p.c. of lead $= 10 \times 0.026 = 0.26\ \Omega$

Test reading, less lead resistance $= 0.34 - 0.26 = 0.08\ \Omega$

This is less than the maximum of $0.1\ \Omega$, so, yes the appliance is satisfactorily earthed, and the test reading can be overruled to 'pass'.

The only problem with this approach is that most portable appliance testers have electronic memory which can be downloaded to software on a PC, which would record 0.54 Ω and a 'fail' status. Unless the instrument or the software includes the facility to include lead resistance, the appliance still fails (something to be said for paper records?).

Having made the above comments, it must be said that only low power appliances with very long cables having small size conductors, cause any problems.

Conducting the earth continuity test

Portable appliance tester

Having conducted the preliminary inspection:

- Plug the appliance into the tester and select, if possible, a suitable current. This will be 1.5 times the fuse rating (if the correct fuse is in place) up to a maximum of 25 A.
- Connect the earth bond lead supplied with the tester, to a suitable earthed point on the appliance. (Remember that just because there is metal, it does not mean that it is connected to earth.) A fixing screw securing the outer casing to a frame is often the best place, rather than the actual casing, which may be enamelled or painted and may contribute to a high resistance reading. If a high reading is obtained, other points on the casing should be tried.
- Start the test, and record the test results.
- Do not touch the appliance during the test.

Figure 5.1 illustrates such a test.

Continuity tester

The method is in general as for the portable appliance tester:

- Zero the instrument.

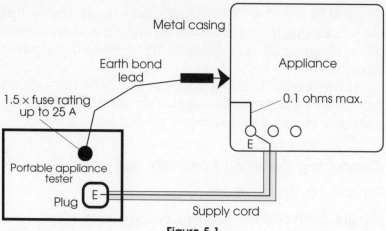

Figure 5.1

- Connect one lead to the earth pin of the plug.
- Connect the other lead to the appliance casing.
- Start the test and record the test results.
- Do not touch the appliance during the test.

Figure 5.2 illustrates such a test.

For the purpose of conducting an earth continuity test using a separate instrument, it would be useful to construct

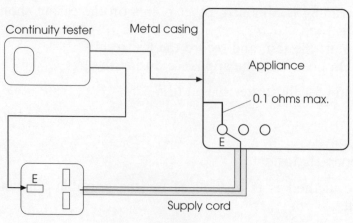

Figure 5.2

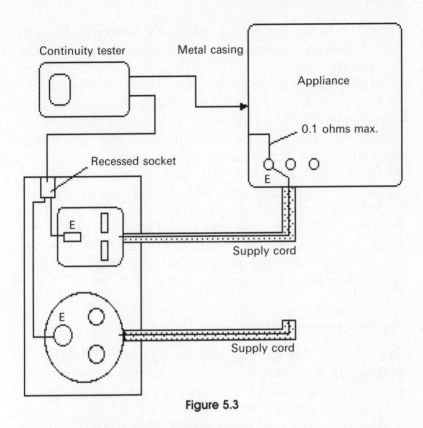

Figure 5.3

a simple means of 'plugging-in' and measuring, rather than trying to make contact with plug pins using clips or probes.

The resourceful tester will make up his/her own aids to testing. Such an aid in this case could be a polypropylene box housing a 13 A and a 110 V socket, with the earth terminals brought out to a metal earth stud suitable for the connection of a test lead (Fig. 5.3).

Again, in the case of testing items of equipment that have to be disconnected from the supply, special test accessories are useful to aid the testing process. Such an accessory would be, for example, a plug, short lead and connector unit, to which a disconnected item could be connected. This is especially useful when using a portable appliance tester, whereas a continuity tester can be connected easily to the exposed protective conductor of the equipment.

Multiway extension sockets and extension leads are to be treated as Class I equipment. However, there is some difficulty in gaining a connection to the earth pin of socket outlets and female plugs. Poking a small screwdriver into the earth socket is not good working practice.

For Class I cord sets, why not use the arrangement shown in Figure 5.3 and add a selection of recessed sockets to house the range of female plugs found on cord sets? All their earth pins would be connected to the earth stud. For extension leads incorporating a socket or sockets, use the earth pin from an old plug, as this is designed to enter the earth pin socket.

Insulation resistance

Realistically, this test can only be carried out on Class I equipment. It is made to ensure that there is no breakdown of insulation between the protective earth and live (phase and neutral) parts of the appliance and its lead.

For Class II items, there are no earthed parts and one test probe would need to be placed at various points on the body of the appliance in order to check the integrity of the casing.

Items that have a cord set, e.g. a kettle, should have the cord set plugged into the appliance and the appliance switch should be in the 'on' position.

There are two tests that can be made, using either the applied voltage method or the earth leakage method.

The applied voltage method

This is conducted using an insulation resistance tester, set on 500 V d.c. The test is made between the phase and

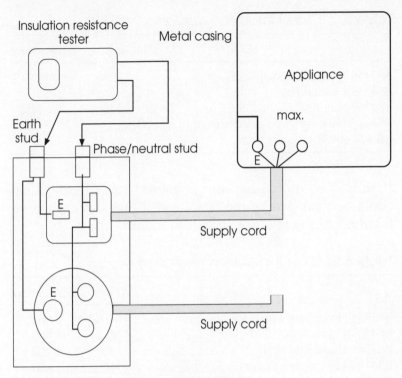

Figure 5.4

neutral **connected together**, and the protective earth. (For three-phase items, all live conductors are connected together.) This is best achieved using the same arrangement as shown in Figure 5.2, but with the addition of a phase/neutral stud connected to the sockets phase and neutral (Fig. 5.4).

Care must be taken when conducting this test to ensure that the appliance is not touched during the process. Also, it should be noted that some items of equipment have filter networks connected across phase and earth terminals and this may lead to unduly low values. The values recorded should not be less than those shown in Tables 5.1 and 5.2.

Table 5.1. Class I insulation resistance values

	As new	In-service
BS3456 Household appliances BS4533 Luminaires BS2769 Hand-held tools BS415 Mains operated electronic equipment BS EN 60950 IT equipment	2 MΩ	0.5 MΩ

It should be noted that any of these items of equipment could be damaged by the 500 V test if they are not manufactured to the relevant British Standard.

Table 5.2. Class II insulation resistance values

	As new	In-service
BS3456 Household appliances	7 MΩ	1 MΩ
BS4533 Luminaires	4 MΩ	1 MΩ
BS2769 Hand-held tools	7 MΩ	1 MΩ
BS415 Mains operated electronic equipment	4 MΩ	1 MΩ
BS EN 60950 IT equipment	7 MΩ	1 MΩ

The earth leakage method

This is achieved using a portable appliance tester that subjects the insulation to a less onerous voltage (usually 250 V) than that delivered by an insulation resistance tester. Here, the leakage current across the insulation is measured, and appliance testers usually set the maximum value at 3.5 mA.

Whichever method is used, there is a chance of pessimistically low values occurring when some heating or cooking appliances are tested. This is usually due to moisture seeping into the insulation of the elements. In this case it is wise to switch such equipment on for a short while to dry the elements out before testing.

NOTE: Many portable appliance testers have the facility to conduct a 'dielectric strength' or 'flash' test, which is basically an insulation resistance test at 1250 V for Class I equipment and 3570 V for Class II. Such voltages could cause damage to insulation and should **not** be carried out for in-service tests.

Functional checks

If testing has been carried out using separate instruments, just switch the equipment to ensure that it is working. If a portable appliance tester is used, there is usually a facility for conducting a 'load test'. The equipment is automatically switched on and the power consumption measured while the item is on load. This is useful as it indicates if the equipment is working to its full capacity, e.g. a 2 kW reading on a 3 kW heater suggests a broken element.

Appendix 1: Basic electrical theory revision

This appendix has been added in order to jog the memory of those who have some electrical background and to offer a basic explanation of theory topics within this book for those relatively new to the subject.

Electrical quantities and units:

Quantity	Symbol	Units
Current	I	Ampere (A)
Voltage	V	Volt (V)
Resistance	R	Ohm (Ω)
Power	P	Watt (W)

Current

This is the flow of electrons in a conductor.

Voltage

This is the electrical pressure causing the current to flow.

Resistance

This is the opposition to the flow of current in a conductor determined by its length, cross sectional area, and temperature.

Power

This is the product of current and voltage, hence $P = I \times V$.

Relationship between voltage, current and resistance

Voltage = Current × Resistance $V = I \times R$ or,
Current = Voltage/Resistance $I = V/R$ or,
Resistance = Voltage/Current $R = V/I$

Common multiples of units

Current I amperes	kA	mA
	kilo-amperes	milli-amperes
	1000 amperes	1/1000 of an ampere
Voltage V volts	kV	mV
	kilovolts	millivolts
	1000 volts	1/1000 of a volt
Resistance R ohms	MΩ	mΩ
	megohms	milli-ohms
	1,000,000 ohms	1/1000 of an ohm
Power P watts	MW	kW
	megawatt	kilowatt
	1,000,000 watts	1000 watts

Resistance in series

These are resistances joined end to end in the form of a

$$R_{total} = R_1 + R_2 + R_3 + R_4$$

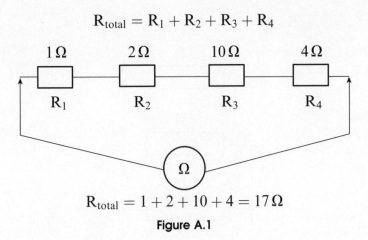

$$R_{total} = 1 + 2 + 10 + 4 = 17\,\Omega$$

Figure A.1

chain. The total resistance increases as more resistances are added (Fig. A.1).

Hence, if a cable length is increased, its resistance will increase in proportion. For example, a 100 m length of conductor has twice the resistance of a 50 m length of the same diameter.

Resistance in parallel

These are resistances joined like the rungs of a ladder. Here the total resistance decreases the more there are (Fig. A.2).

The insulation between conductors is in fact countless millions of very high value resistances in parallel. Hence an increase in cable length results in a decrease in insulation resistance. This value is measured in millions of ohms, i.e. megohms ($M\Omega$).

The overall resistance of two or more conductors will also decrease if they are connected in parallel (Fig. A.3).

The total resistance will be half of either one and would be the same as the resistance of a $2\,mm^2$ conductor. Hence resistance decreases if conductor cross sectional area increases.

$$1/R_{total} = 1/R_1 + 1/R_2 + 1/R_3 + 1/R_4$$

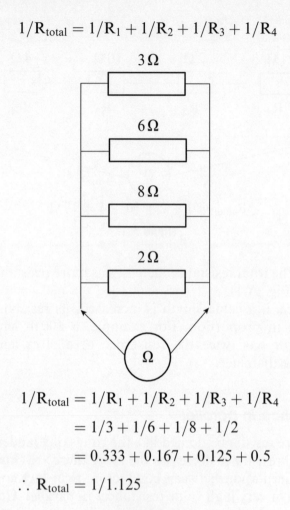

$$1/R_{total} = 1/R_1 + 1/R_2 + 1/R_3 + 1/R_4$$
$$= 1/3 + 1/6 + 1/8 + 1/2$$
$$= 0.333 + 0.167 + 0.125 + 0.5$$
$$\therefore R_{total} = 1/1.125$$

Figure A.2

Example

If the resistance of a $1.0\,mm^2$ conductor is $19.5\,m\Omega/m$, what would be the resistance of:

1   85 m of $1.0\,mm^2$ conductor
2   1 m of $6.0\,mm^2$ conductor

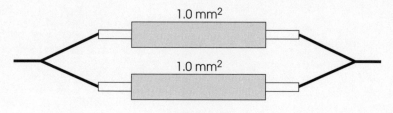

Figure A.3

3   25 m of 4.0 mm^2 conductor
4   12 m of 0.75 mm^2 conductor

Answers

1   1.0 mm^2 is 19.5 mΩ/m, so, 85 m would be 19.5 × 85/1000 = 1.65 Ω
2   A 6.0 mm^2 conductor would have a resistance 6 times less than a 1.0 mm^2 conductor, i.e. 19.5/6 = 3.35 mΩ
3   25 m of 4.0 mm^2 would be 19.5 × 25/4 × 1000 = 0.12 Ω
4   12 m of 0.75 mm^2 would be 19.5 × 12/0.75 × 1000 = 0.312 Ω

Appendix 2
Sample 2377
questions

The management of electrical equipment maintenance

1 Which one of the following is a statutory document?

   a. A British Standard
   b. IEE Wiring Regulations
   c. IEE Codes of Practice
   d. Electricity at Work Regulations.

2 Which one of the following regulations states: *'Every employer shall make a suitable and sufficient assessment of the risk to the health and safety of his employees and to persons not in his employment'*?

   a. The Electricity Supply Regulations
   b. The Electricity at Work Regulations
   c. The Provision and Use of Work Equipment Regulations
   d. The Management of Health and Safety at Work Regulations.

3   Certain sections of The Health and Safety at Work Regulations put a duty of care upon:

    a. employees only
    b. employers only
    c. both employees and the general public
    d. both employers and employees.

4   Which one of the following regulations state: *'As may be necessary to prevent danger, all systems shall be maintained so as to prevent, so far as is reasonably practicable, such danger'*?

    a. The Electricity at Work Regulations
    b. The IEE Wiring Regulations
    c. The Provision and Use of Work Equipment Regulations
    d. The Management of Health and Safety at Work Regulations.

5   The scope of legislation of inspection and testing of electrical equipment extends to distribution systems up to:

    a. 230 V
    b. 400 V
    c. 11 kV
    d. 400 kV.

6   The Code of Practice for In-service Inspection and Testing of Electrical Equipment does **not** apply to:

    a. shops
    b. offices
    c. caravan sites
    d. petrol station forecourts.

7   The safety and proper functioning of certain portable appliances and equipment depends on the integrity of the fixed installation. Requirements for the inspecting and testing of fixed installations are given in:

    a. BS2754
    b. BS7671
    c. BSEN60947
    d. BSEN60898.

8   Transportable equipment is sometimes called:

    a. hand-held appliance or equipment
    b. stationary equipment or appliance
    c. moveable equipment
    d. portable appliance.

9   An electric kettle is classified as a:

    a. portable appliance
    b. moveable equipment
    c. hand-held appliance
    d. equipment for 'building in'.

10   Which one of the following domestic electrical appliances may be regarded as an item of stationary equipment?

    a. A bathroom heater
    b. A visual display unit
    c. A washing machine
    d. A built-in electric cooker.

11   A portable appliance that is supplied by a flexible cord incorporating a protective conductor, is classified as:

    a. Class I
    b. double insulated
    c. metal clad Class II
    d. Class III.

12 Stationary equipment/appliances are defined as not being provided with a carrying handle and have a mass exceeding:

   a. 10 kg
   b. 12 kg
   c. 15 kg
   d. 18 kg.

13 A freezer is classified as a:

   a. stationary appliance or equipment
   b. hand-held appliance or equipment
   c. moveable equipment
   d. portable appliance.

14 A BS 3535 safety isolating transformer having a voltage not exceeding 50 V is used to supply certain equipment. The class of such equipment is

   a. Class 0
   b. Class I
   c. Class II
   d. Class III.

15 Which size of the following three-core extension leads is too large for a standard 13 A plug?

   a. $2.5\,\text{mm}^2$
   b. $1.5\,\text{mm}^2$
   c. $1.25\,\text{mm}^2$
   d. $1.00\,\text{mm}^2$.

16 Which one of the following arrangements would **not** meet the requirements of the IEE Code of Practice?

   a. Class I equipment supplied by a $1.5\,\text{mm}^2$ three-core extension lead connected into a 13 A three-pin socket outlet.

b. Class II equipment supplied by a $1.5\,mm^2$ two-core extension lead connected into a 13 A three-pin socket outlet.

c. Class I equipment supplied by a $2.5\,mm^2$ three-core extension lead connected into a **BS EN** 60309-2 socket outlet.

d. Class III equipment supplied by a two-core flexible cord connected into the secondary of an isolating transformer supplying SELV lighting equipment.

17 Which one of the following size and length extension leads should be used in conjunction with an r.c.d. used for supplementary protection?

a. $1.5\,mm^2$, 10 m long
b. $1.5\,mm^2$, 15 m long
c. $2.5\,mm^2$, 20 m long
d. $2.5\,mm^2$, 30 m long?

18 During the inspection and testing process, which of the following is not required:

a. Preliminary inspection
b. Earth continuity tests (for Class I equipment)
c. Insulation testing
d. Earth continuity test on Class II equipment?

19 Which one of the following would **not** be conducted during routine inspection and testing of appliances:

a. Preliminary inspection
b. Earth continuity tests
c. Type testing
d. Functional checks?

20 When performing in-service testing on Class I equipment, which one of the following is **not** required:

   a. Type testing to a British Standard
   b. Earth continuity test
   c. Insulation testing
   d. Functional checks?

21 Details of which of the following must be recorded when carrying out a safety check on an electrical appliance:

   a. Manufacturer's name and address
   b. Combined inspection and test
   c. User check revealing no damage to equipment
   d. Applicable British Standards?

22 Which one of the following will **not** affect the frequency of inspection and testing for an electrical appliance:

   a. The integrity of the fixed electrical installation
   b. Environment in which it is to be used
   c. The user
   d. The equipment class?

23 Recorded testing but not inspecting of equipment may be omitted if the:

   a. equipment is of Class I construction and in a low-risk area
   b. equipment is of Class II construction and in a low-risk area
   c. user of the equipment reports damage as and when it becomes evident
   d. equipment is a hand-held appliance.

24 The table of suggested frequency of inspection and testing for electrical equipment gives details of:

    a. the forms required for such testing
    b. maximum and minimum values of test results
    c. the required sequence of visual checks to be made
    d. types of premises within which electrical equipment is operated and user check requirements.

25 The suggested initial frequency for a formal visual inspection of a hand-held Class II electric iron in a hotel is:

    a. one month
    b. six months
    c. 12 months
    d. 24 months.

26 The suggested frequency for user checks for children's rides in a fairground is:

    a. weekly
    b. monthly
    c. daily
    d. 12 months.

27 Which one of the following tests should **not** be applied routinely to equipment:

    a. Earth continuity
    b. Insulation resistance
    c. Polarity
    d. Dielectric strength?

28 The first electrical test to be applied to Class I equipment is:

    a. insulation resistance
    b. earth continuity

   c. dielectric strength

   d. polarity.

29 When information regarding test procedures is unavailable from the manufacturer or supplier of IT equipment which one of the following electrical tests should **not** be undertaken?

   a. Earth continuity

   b. Polarity

   c. Functional

   d. Insulation.

30 The purpose of an equipment register is to ensure:

   a. compliance with the Electricity at Work Regulations

   b. that maintenance procedures are recorded

   c. the frequency of inspection and test is reviewed

   d. inspection and testing is performed.

31 Identification of all electrical equipment within a duty holder's control is required in order to produce:

   a. 'pass' safety check equipment label

   b. faulty equipment register

   c. equipment register

   d. repair register.

32 Which one of the following items of information is **not** required on an inspection and test label?

   a. An indication of whether the equipment has passed or failed the safety tests

   b. Details of previous test results

   c. Date at time of testing

   d. Appliance or equipment number.

33 All electrical equipment should be marked with a unique serial number to aid:

a. disconnection
b. identification
c. risk assessment
d. interpretation of test results.

34 Information provided for equipment which requires routine inspection and/or testing should consist of:

a. an indelible bar-code system
b. an identification code to enable the equipment to be uniquely identifiable
c. operating instructions regarding the test equipment
d. an indication of the results which may be expected during inspections and/or tests.

35 Which one of the following is **not** required to be tested within the scope of the IEE Code of Practice?

a. Fixed equipment
b. Fixed installations
c. Electrical tools
d. Portable appliances.

36 The Memorandum of Guidance on the Electricity at Work Regulations 1989 advises that equipment records:

a. should be kept throughout the working life of the equipment
b. only be kept where the equipment is used in high-risk areas
c. are not required where the equipment is used in low-risk areas

    d. are not required if the equipment is fed from a 110 V safety supply.

37 Records of all maintenance activities relating to electrical appliances must be kept, including details of the:

    a. initial cost
    b. procurement of equipment
    c. estimated replacement date
    d. estimated replacement cost.

38 The person responsible for carrying out an inspection and test on an appliance should have made available to them:

    a. a list of all the users of equipment
    b. a copy of the Electricity at Work Regulations
    c. a copy of the Health and Safety at Work Act
    d. previous inspection and test results.

39 Which voltage must be used when carrying out an insulation resistance test on a Class I toaster?

    a. 3750 V a.c.
    b. 500 V d.c.
    c. 1000 V d.c.
    d. 500 V a.c.

40 An insulation resistance tester should be capable of:

    a. delivering a minimum voltage of 1000 V d.c. to the load
    b. testing the continuity of an earthing circuit
    c. delivering a maximum voltage of 25 A through the load
    d. maintaining the test voltage required across the load.

41 Where a user check reveals damage to equipment, it must be reported to:

a. the equipment manufacturer
b. the Health and Safety Inspectorate
c. a responsible person
d. a manager of an inspection and test organisation.

42 The manager of an inspection and test organisation should be able to:

a. repair faulty electrical equipment
b. instruct untrained persons in the use of portable appliance testers
c. know their legal responsibilities under the Electricity at Work Regulations
d. demonstrate competence in the use of appliance testers.

43 Which one of the following is outside the scope of the IEE Code of Practice for Inspection and Testing of in-Service Electrical equipment?

a. Those who inspect and test
b. The user of electrical appliances
c. Managers of the inspection and test organisation
d. The hirer of electrical portable appliances and equipment.

44 Earth continuity testing may in certain circumstances be carried out by means of:

a. a low resistance ohmmeter
b. an insulation resistance tester
c. a bell set and battery
d. an instrument complying with BS EN 60309.

45 Test leads and probes used to measure voltages over 50 V a.c. and 100 V d.c. should comply with:

a. BS 7671
b. Health and Safety Executive Guidance Note GS 38
c. BS 5490 Specification for classification of Protection
d. IEC Publication 479.

Appendix 3
Sample 2377 questions

Inspection and testing of electrical equipment

1   Where protection against electric shock from equipment is provided using a protective conductor in the fixed wiring, the equipment classification would be:

   a. Class I
   b. Class II
   c. Class III
   d. Class 0.

2   A safety isolating transformer for Class III equipment must conform to:

   a. BS 3456
   b. BS 3535
   c. BS 4533
   d. BS 5458.

3   A substantially continuous metal enclosure associated with Class II equipment, would be classified as:

   a. insulation encased
   b. isolation encased
   c. metal-cased
   d. metal insulated.

4   There is no provision for protective earthing or reliance upon installation conditions for which one of the following equipment?

   a. Class I
   b. Class II
   c. Class III
   d. Class 01.

5   Which one of the following is the Class III construction mark?

   a.   b.

   c.   d.

6   Which one of the following is the Class II construction mark?

   a.   b.

   c.   d.

7   The suggested initial frequency of user checks, relevant to a children's ride sited in the entrance of a large store, could well be:

   a. daily
   b. monthly
   c. every 3 months
   d. every 6 months.

8   Which voltage should be applied when conducting an insulation resistance test on an electrical appliance?

    a. 230 V a.c.
    b. 230 V d.c.
    c. 500 V a.c.
    d. 500 V d.c.

9   User checks of stationary equipment installed in industrial premises should be conducted:

    a. before use
    b. daily
    c. weekly
    d. monthly.

10 When assessing the level of safety of an electrical appliance, the most important check would be:

    a. visual inspection
    b. flash testing
    c. earth linkage current
    d. the minimum acceptable values of insulation resistance.

11 Which one of the following checks should the user be competent to undertake?

    a. Combined inspection and testing
    b. Tests using a portable appliance tester
    c. Visual inspection of the flexible lead and plug fitted to an appliance
    d. Formal visual inspection.

12 A user of equipment should be competent to inspect:

    a. terminal screws
    b. socket outlets
    c. equipment fuses
    d. protective conductors.

13 During a formal visual inspection it should be confirmed that the equipment is being operated:

a. at the correct voltage
b. by a skilled person
c. by an instructed person
d. in accordance with manufacturer's instructions.

14 If a standard 13 A plug became overheated the most likely cause would be:

a. a loose connection at one or more of the terminals
b. reversed polarity of the cable conductors
c. inadequate earthing connections
d. the use of an incorrectly rated cartridge fuse.

15 Before isolating the supply to a computer system, the inspector should ensure that:

a. all recent data is downloaded and saved
b. permission is obtained from the equipment user
c. permission is obtained from the responsible person
d. any static electricity is discharged.

16 When conducting a combined inspection and test, the visual inspection should determine:

a. the nature of the tests to be conducted when the equipment is not allowed to be disconnected from the supply
b. whether all 13 A fused plugs fitted to portable appliances should be to BS 4343 or BS EN 6039-2
c. if any disconnected optical fibre cabling should have exposed ends dipped in a scaling solvent in order to exclude moisture

d. whether the equipment and/or its flexible cord has suffered any physical damage.

17 When conducting insulation resistance tests on new household electrical appliances to BS 3456 Class I insulation, the minimum value would be:

a. 0.5 megohm
b. 1.0 megohm
c. 2 megohm
d. 7 megohm.

18 Which test should be carried out on low voltage electronic equipment within a computer suite?

a. Earth continuity test at 12 V
b. Insulation resistance test using the earth leakage method
c. Flash test
d. Functional test with equipment on load.

19 The maximum permitted length of a $1.25\,mm^2$ extension lead fitted with a standard 13 A plug should not exceed:

a. 6 m
b. 12 m
c. 15 m
d. 25 m.

20 Which one of the following would **not** be applicable for a test on a two-core cord set?

a. Visual inspection
b. Earth continuity test
c. Polarity check
d. Insulation resistance test.

21 An ohmmeter used to measure the resistance of an earth continuity conductor, must be capable of producing a short-circuit current between:

    a. 2 and 10 mA
    b. 10 and 20 mA
    c. 20 and 200 mA
    d. 200 and 500 mA.

22 An insulation resistance test of a Class I household portable appliance to BS 3456 is to be carried out using the earth leakage method. The maximum acceptable value is:

    a. 0.25 mA
    b. 0.5 mA
    c. 0.75 mA
    d. 1 mA.

23 A BS 2769 Class II portable electric drill has been refurbished and classified 'as-new'. The minimum acceptable value of insulation resistance when tested would be:

    a. 0.5 MΩ
    b. 1.5 MΩ
    c. 2.5 MΩ
    d. 7.5 MΩ.

24 Which one of the following is **not** required on an equipment inspection and testing label?

    a. Date of check
    b. Identification number
    c. Age of equipment
    d. Re-test period.

25 Equipment found to be faulty must not be used but must be:

 a. labelled and reported
 b. labelled and withdrawn from service
 c. reported and withdrawn from service
 d. labelled, reported and withdrawn from service.

26 A 2-cord set is to be tested separately from the appliance. Which one of the following is **not** applicable?

 a. Visual inspection
 b. Earth continuity
 c. Insulation
 d. Polarity.

27 The length of a $1.5\,\text{mm}^2$ extension lead should not exceed:

 a. 10 m
 b. 12 m
 c. 15 m
 d. 25 m.

28 A $1.25\,\text{mm}^2$ extension lead 15 m long should be protected by a:

 a. 30 mA residual current device
 b. semi-enclosed fuse
 c. miniature circuit breaker
 d. cartridge fuse.

29 IT equipment which is **not** constructed to BS EN 60950 may be damaged by an applied voltage insulation resistance test. The test that should replace it is:

 a. a polarity test
 b. a dielectric strength test

c. a continuity test

d. an earth leakage test.

30 Equipment with an earth leakage current designed to exceed 3.5 mA shall:

a. have a label permanently fixed indicating the value of leakage current

b. have internal protective conductors of not less than $0.5\,mm^2$ c.s.a.

c. be permanently wired or supplied by a plug and socket to BS 4343 (BS EN 6030-2)

d. only be used in industrial situations.

Appendix 4
Answers to test questions

Answers to Appendix 2 questions

1. d	2. d	3. d	4. a	5. d
6. d	7. b	8. c	9. c	10. c
11. a	12. d	13. a	14. d	15. a
16. b	17. d	18. d	19. c	20. a
21. c	22. a	23. b	24. d	25. b
26. d	27. d	28. b	29. d	30. c
31. c	32. b	33. b	34. b	35. b
36. a	37. b	38. d	39. b	40. d
41. c	42. c	43. d	44. a	45. b

Answers to Appendix 3 questions

1. a	2. b	3. c	4. b	5. d
6. b	7. a	8. d	9. b	10. a
11. c	12. b	13. d	14. a	15. c
16. d	17. c	18. b	19. b	20. b
21. c	22. c	23. d	24. c	25. d
26. b	27. c	28. a	29. d	30. c

Index

Index

L

Legislation, 1

M

Management of The Health &
  Safety at Work Act, 1

P

Plugs and Sockets Regulations,
  26
Portable Appliance Tester, 28,
  31
Power, 40
Preliminary inspection, 28
Provision & use of work
  equipment act, 1, 2

R

Resistance, 40
Resistance in parallel, 41
Resistance in series, 40
Responsible person, 7, 8, 27

T

Testing, 27, 28
Testing extension leads, 33

U

Units, 37
User checks, 23

V

Voltage, 39